What Survives

Dimension, Decoherence and the Birth of Spacetime

Brian Capleton PhD

What Survives : Dimension, Decoherence and the Birth of Spacetime

Amarilli Books
ISBN: 978-1-917403-24-5

Some of the material on the IIP-VGF mathematical framework in this book appears also in *The Infinite Iteration Principle - Vol. 6 : Mathematical Foundations*

The IIP-VGF framework presented here has been developed from earlier exploratory work on the infinite iteration principle outside the domain of standard mathematics (*Structures and Morphisms,* 2017 and *Objects and Structures,* 2022). The present IIP-VGF framework is based on standard mathematics and has been made possible through intensive use of artificial intelligence which until now was not available with the necessary power.

- *The Infinite Iteration Principle - Vol. 1 : Physics*
- *The Infinite Iteration Principle - Vol. 2 : Evolution and Intelligence*
- *The Infinite Iteration Principle - Vol. 3 : Brain and Mind*
- *The Infinite Iteration Principle - Vol. 4 : The Partition in Quantum Decoherence*
- *The Infinite Iteration Principle - Vol. 5 : Key Areas*
- *The Infinite Iteration Principle - Vol. 6 : Mathematical Foundations*

Contents

Introduction

In the IIP-VGF framework the primitive, generative principle of nature, is simply iteration. This iteration principle is a raw principle, and pre-mathematical. Persistence under iteration then gives rise to relations that survive and can be treated as structure. In this sense, the framework is neither realist about objects nor realist about structure in the traditional way. It is instead realist about generativity: what is fundamental is the iterative activity from which both objects and structures arise.

Mathematics and structural description then enter retrospectively, as the language of those patterns that survive. Structural realism correctly identifies the durability of structure across theory change, but the IIP–VGF framework explains why such structures are durable in the first place — because as we can now understand them using mathematics, they are the fixed points of iteration under loss of fidelity. But in the IIP-VGF framework nature is not mathematical and then physical; it is generative first, and only later describable mathematically once something has survived.

What we shall see in this book is that at the deepest level, the fundamental principles and constants of nature do not describe substances, forces, or even laws in the traditional sense. They describe conditions of survival.

What we call physical laws, conserved quantities, dimensions, probabilities, and constants are best understood as invariants of persistence — structures that remain stable under repeated interaction while others are erased.

In the IIP–VGF framework, the generative principle underlying this is iteration itself. Iteration produces variation, interference, and loss of fidelity; only those configurations that can stabilise under repeated application endure. The universe does not begin with objects and then ask how they behave; it begins with iteration, and only later acquires objects as fixed points of that process. Nature, in this sense, is not defined by what is possible, but by what survives.

Modern physics rests on two pillars whose coexistence remains uneasy. Quantum theory describes the microscopic world in terms of superposition, entanglement, and probability amplitudes. Relativity describes the large-scale structure of reality as a smooth spacetime endowed with geometry, dimensionality, and causal order. Both frameworks are extraordinarily successful within their domains, yet both presuppose classical structures that they do not themselves explain.

Quantum theory presumes a classical spacetime in which measurements occur; relativity presumes a classical geometry whose dimensionality is fixed from the outset. The question of how classical spacetime itself comes into being therefore lies at the fault line between our most fundamental theories.

Over the past decades, decoherence theory and quantum Darwinism have substantially clarified how classical measurement outcomes emerge from quantum dynamics. They have shown that classicality is not imposed by an external observer, but arises through iterative interaction, redundancy, and the selective stabilisation of certain structures over others. What survives is what can be copied, reinforced, and rendered objective. Yet these advances are typically formulated within an already-given spacetime, leaving open a deeper question: if decoherence explains the emergence of classical outcomes, can the same principle explain the emergence of classical space itself?

This book argues that it can.

The central claim developed here is that classical spacetime, spatial dimensionality, and measurement outcomes are products of the same generative mechanism. That mechanism is not collapse, not observation, and not a special cosmological initial condition, but a universal process of iterative stabilisation under loss of fidelity. Classical reality arises because certain structures persist under repeated interaction while others do not. In this sense, reality is not what exists once, but what survives.

To articulate this claim, the book introduces the Infinite Iteration Principle and the Vast Generative Field (IIP–VGF), a minimal formal framework in which iteration is taken as primitive and structure arises through closure. Within this framework, physical laws, mathematical invariants, and spacetime geometry are treated as fixed points of a generative process rather than as axiomatic inputs. Stability and fidelity form a universal trade-off: precision is sacrificed so that structure can persist, propagate, and become objective.

The argument proceeds by following the earliest physical fidelity cascade. From a pre-geometric regime, stabilised locality emerges. From locality, spacetime appears as the first coherence-preserving envelope. Within spacetime, conserved dynamical structure gives rise to energy, which in turn supports particles as localised, repeatable excitation-closures. Spatial dimensionality is shown not to be a logical necessity but an attractor: it is what survives coarse-graining and redundancy. Three dimensions are not derived; they are selected.

Decoherence is then re-examined from this perspective. Pointer bases and spatial regions are shown to be instances of the same algebraic phenomenon: idempotent decompositions of a stabilised structure. Measurement outcomes and spatial localisation arise by the same logic, at different levels of the fidelity cascade. The book concludes by clarifying the status of probability and the Born rule. Probability is not introduced as an arbitrary axiom, but as a measure of stabilisation: the squared norm appears because it is the unique invariant compatible with quadratic self-composition and redundancy. In this sense, probability and dimensionality occupy analogous roles as measures associated with what survives.

How to Read the Mathematics in This Book

The mathematics in the chapters that follow is deliberately spare and structural. It is not presented as a sequence of axioms leading to theorems, nor as a proposal for new dynamical laws. Instead, mathematical objects — idempotents, fixed points, algebras, coarse-graining flows, scaling exponents — are used as stability markers: they name what persists under iteration and loss of fidelity.

The mathematics used here should not be read as a claim about what nature **is**, but as a disciplined way of describing **what nature retains**. In other words, what has survived. The IIP–VGF framework does not assume that the universe is fundamentally mathematical. Mathematics appears here neither as an ontological substrate nor as an abstract realm from which the physical world is instantiated.

Rather, it is employed as the most precise language available for articulating stable distinctions once such distinctions have emerged.

The generative activity posited at the base of the framework — iteration prior to structure — is pre-mathematical. Before closure, there are no objects, quantities, domains, or relations in the formal sense. Number, identity, and relation arise only when certain patterns persist under repeated iteration. Mathematics becomes applicable after this point, because closure produces invariants, composability, and repeatability. In this sense, mathematics is not the origin of structure, but its afterimage.

Nevertheless, once stabilisation has occurred, mathematics can be used retrospectively to formalise the generative process itself. Fixed points, idempotents, algebras, coarse-graining flows, and scaling exponents do not posit new entities; they name what survives iteration under loss of fidelity. The formalism reconstructs generativity from within its outcomes, expressing persistence, redundancy, and stability in invariant terms. This is why mathematics appears throughout the book even when discussing pre-geometric or pre-classical regimes: it is not because those regimes are mathematical in themselves, but **because their surviving traces admit mathematical description**.

Equations in this book should therefore be read operationally rather than representationally. They do not assert that nature obeys exact mathematical laws at all scales, nor that the universe instantiates a particular formal structure. Instead, they indicate that certain patterns are stable under further iteration. When a quantity is conserved, it marks a closure. When an algebra becomes commutative, it marks a geometric regime. When a quadratic form appears, it marks redundancy and stabilisation. The **mathematics functions as a map of survivability**, not as a catalogue of fundamental ingredients.

Readers should resist the temptation to ask, at each step, "What is being postulated here?" The more appropriate question is: What is being stabilised? The formalism is cumulative and comparative rather than axiomatic. Each mathematical structure gains its meaning from its position in the cascade — from pre-geometric iteration, through locality and spacetime, to decoherence, measurement, and probability. The **mathematics does not ground the framework; it tracks its fixed points**.

Approached in this way, the formal material serves a unifying purpose. It shows why the same stabilising logic reappears across domains that are usually treated as unrelated: spatial dimensionality, classical geometry, measurement outcomes, probability weights, and cosmological acceleration. These are not separate mathematical facts imposed on nature, but different expressions of the same principle — that what becomes the actuality of nature is what survives.

The IIP-VGF Framework

Central to the IIP–VGF framework is the Stability–Fidelity relation. In physics, it first manifests in quantum decoherence, where stable, objective measurement outcomes arise through the loss of fidelity to underlying quantum coherence.

In physical terms, the Stability–Fidelity relation describes the trade-off that arises when a system becomes robust through interaction with its environment. At the quantum level, isolated systems can retain high fidelity to their underlying coherent state, but such coherence is fragile and cannot persist once the system is repeatedly and redundantly coupled to external degrees of freedom.

Through decoherence, specific relational patterns between the system and its environment stabilise, producing persistent, effectively classical structures — such as definite positions, trajectories, or pointer states — that can be repeatedly recorded and agreed upon.

This stabilisation increases physical objectivity and robustness precisely because information about the system is redundantly distributed across the environment. However, the cost of this robustness is a loss of fidelity to the original coherent dynamics: phase relations and generative superpositions are irreversibly thinned.

In this way, the physical world acquires stability and causal regularity only by relinquishing direct access to its deeper coherent generative structure, exemplifying the universal trade-off between stability and fidelity.

Within the IIP–VGF framework, quantum decoherence is understood not as a domain-specific mechanism peculiar to quantum theory, but as the first salient physical expression of a more general principle: the Stability–Fidelity relation governing all stabilised structures.

Decoherence exemplifies how generative coherence, when repeatedly iterated and coupled to an expanding environment, stabilises by trading *fidelity to its underlying dynamics*, for robustness, redundancy, and persistence. The IIP–VGF framework extends this logic beyond quantum systems by treating decoherence as a universal mode of closure, through which any coherent generative process — physical, biological, cognitive, or conceptual — produces stable, objective forms at the cost of thinning its connection to deeper coherence.

In this sense, quantum decoherence is not merely a transition from quantum to classical behaviour, but a specific instance of a broader evolutionary pattern by which structure itself arises: **stability is achieved only by relinquishing direct fidelity to the generative field from which it emerges**.

The Initial Proposition

Formally, if iteration is represented as the operator ⊙ acting on itself, then the VGF represents the closure of this process under infinite composition:

$$\odot \rightarrow \odot(\odot) \rightarrow \odot(\odot(\odot)) \rightarrow \ \dots \ \rightarrow \ lim \rightarrow \infty \ \odot(\odot)$$

Each stabilised form within this infinite composition corresponds to an idempotent or fixed-point structure, denoted generically as

$$E_i = \mathbb{I}(E_i)$$

In a Nutshell

Formally, let $\mathcal{I}$ denote the Infinite Iteration Principle as an endo-operator on a proto-domain $\mathcal{X}$, and define the VGF as the minimal structure $\mathcal{V}$ that is closed under unbounded composition of $\mathcal{I}$:

$$\mathcal{V} = \overline{\{\mathcal{I}^n(x) \mid x \in \mathcal{X},\ n \in \mathbb{N} \cup \{\infty\}\}}.$$

A closure is realised when an iterative trajectory stabilises into an idempotent structure

$$e \in \mathscr{V}, \qquad e \circ e = e,$$

together with an associated functorial envelope

$$F_e : \mathscr{E}_e \to \mathscr{E}_e$$

whose redundant action renders e dynamically self-supporting (environmental stabilisation).

Prior to such stabilisation, candidate idempotents exist only as mutually entangled proto-morphisms within $\mathscr{V}$; they are not yet orthogonalised as distinct closures.

In this sense, $\mathscr{V}$ may be characterised as a field of virtual idempotents — a superposition of unrealised closures — which become articulate only when a functorial envelope $\mathscr{E}_e$ yields a decohered, context-specific fixed point. Each realised closure e thereby induces a new envelope that acts as a higher-level environment for further iteration, such that $\mathscr{V}$ remains globally open while being locally partitioned by recursively generated idempotent refinements.

Categories

Categorically, let $\mathscr{C}_{\mathscr{I}}$ be the category generated by iterative trajectories of $\mathscr{I}$ on $\mathscr{X}$, and let $\mathrm{Kar}(\mathscr{C}_{\mathscr{I}})$ denote its idempotent completion (Karoubi envelope).

Objects of $\mathrm{Kar}(\mathscr{C}_{\mathscr{I}})$ are pairs (A, e) with $A \in \mathrm{Ob}(\mathscr{C}_{\mathscr{I}})$ and $e : A \to A$ idempotent, $e^2 = e$, and morphisms are those in $\mathscr{C}_{\mathscr{I}}$ that respect the splittings of idempotents. In this setting, realised closures correspond to split idempotents (A, e) equipped with stabilising endo-functors

$$\longrightarrow F_{(A,e)} : \mathscr{E}_{(A,e)} \to \mathscr{E}_{(A,e)'} \longrightarrow$$

where $\mathscr{E}_{(A,e)}$ is the functorial environment category encoding the redundancy that renders (A, e) dynamically robust. The pre-closure VGF corresponds, not to $\mathrm{Kar}(\mathscr{C}_{\mathscr{I}})$itself, but to $\mathscr{C}_{\mathscr{I}}$ before idempotent splitting, in which idempotents are present only as unsplit endomorphisms and hence as "virtual closures".

In this precise sense, the VGF can be regarded as a pre-Karoubi superposition of closures: a field of unsplit idempotents whose decohered realisation is given by passage to $\mathrm{Kar}(\mathscr{C}_{\mathscr{I}})$ together with the assignment of stabilising functorial envelopes $\mathscr{E}_{(A,e)}$. Each such realised object (A, e) then functions as a new environmental node in a higher-level categorical recursion of the VGF.

Interpretation

The equation we met above:

$$E_i = \mathbb{I}(E_i)$$

formalises stability within the infinite iterative field: It defines each E_i as an idempotent or fixed-point structure — an entity that reproduces itself under the universal act of iteration.

In physics, such fixed points correspond to pointer states or stable equilibria emerging through decoherence, where repeated environmental interaction leaves the observable unchanged.

In biological evolution, they represent species or morphological attractors — configurations that persist through adaptive cycles by maintaining functional coherence under environmental iteration.

In cognition, they appear as conceptual invariants or memory schemas — stable representations that remain recognisable despite recursive modification through perception and recall.

Across all domains, $E_i = \mathbb{I}(E_i)$ expresses the defining feature of stability under transformation: the system has become its own iteration, conserving relational coherence within the vast generative field even as fidelity to its generative origin has been thinned.

These idempotent stabilisations together constitute the structured fabric of the VGF, through which the infinite iteration process becomes manifest as the coherent order in and structure of the discriminatory observable world.

Such idempotents form the **attractors** of the VGF — self-consistent loops of recursion.

The quantum domain can thus be considered as arising in this way. When quantum *decoherence* is introduced, these attractors become redundantly stabilised, forming the observable objects of the classical domain.

Thus, every measurable entity — whether physical (a particle), mathematical (a number), or even cognitive (a concept) — is a stabilised encoding of the recursion of recursion.

This gives rise to the **Stability–Fidelity Law**: fidelity to the generative recursion is lost as stability (redundancy) is gained.

Decoherence, in this context, is the contraction of recursive self-reference into finite, redundant representation. Quantum decoherence is itself then a particular representation of the more general decoherence principle in the VGF, of stabilisation and redundancy at the expense of fidelity to an underlying iterative dynamic.

Idempotents

Within the IIP–VGF framework, each fundamental physical theory corresponds to a stabilisation of the generative dynamics: formally, an idempotent

$$E_i = \square(E_i)$$

representing a closure of the underlying IIP into a domain of robust, calculable structure. An idempotent E_i selects a consistent subspace of the Vast Generative Field (VGF) within which coherence has thinned into stable, reproducible patterns.

Quantum mechanics and general relativity arise as distinct idempotent closures:

- Quantum mechanics corresponds to an idempotent E_Q that stabilises Hilbert-space structure, superposition, and unitary evolution;
- Relativity corresponds to an idempotent E_R that stabilises Lorentz geometry, local causality, and smooth spacetime manifolds.

Each closure is internally self-consistent —

$$E_Q^2 = E_Q, \qquad E_R^2 = E_R,$$

and each yields high Stability within its own Fidelity domain.

The trouble appears when one attempts to impose both stabilisations simultaneously. In VGF terms, the two idempotents do not commute:

$$E_Q E_R \neq E_R E_Q.$$

The attempted composite

$$E_{QR} = E_Q E_R$$

fails to produce a legitimate idempotent. It is neither stable nor closed under iteration:

$$E_{QR}^2 \neq E_{QR}.$$

This non-closure manifests mathematically as the familiar series of divergences — non-renormalisable integrals, infinite self-energies, and breakdowns of locality. These infinities are therefore not accidental. They are the algebraic trace of the fact that the two stabilisations carve the VGF along incompatible boundaries. Their domain assumptions, symmetry constraints, and redundancy structures are not jointly satisfiable.

In short: the "infinities problem" arises because

$$E_Q \text{ and } E_R$$

are mutually non-coherent idempotents of the generative field. A genuine quantum theory of gravity requires a new idempotent

$$E_* = \mathbb{I}(E_*)$$

whose closure contains both quantum and relativistic regimes as derived, lower-fidelity stabilisations:

$$E_Q = E_* E_Q E_*, \qquad E_R = E_* E_R E_* .$$

Thus the divergences are the mathematical symptom that we are attempting to apply two incompatible stabilisers of the VGF at once, rather than operating within a single, higher-order stabilisation in which they can be consistently unified.

The Canonical Form of the VGF Dynamics

The dynamics of the Vast Generative Field (VGF) are captured in their most compact and universal expression by the quadratic tensor recursion map

$$g(\mathscr{K}) = \alpha\,\mathscr{K}^2 + \beta\,\mathscr{K} + \gamma\,I,$$

where $\mathscr{K}$ denotes the generative operator acting on itself, and α, β, γ are scalar or tensorial coefficients representing the coupling, persistence, and invariance of the recursion respectively.

This form is canonical because it expresses, in the most economical algebraic language, the three essential processes that any self-iterating field must undergo: nonlinear self-coupling, linear continuity, and invariant grounding.

- Nonlinear self-coupling ($\alpha\mathscr{K}^2$) — Every generative system is defined by feedback: it acts on the results of its own prior action. The quadratic term encodes this reflexivity. It represents the mutual influence of the field's internal and external aspects—the "system" and its "environment"—and is therefore the algebraic expression of iteration itself. Without a quadratic self-coupling, the field would evolve linearly and could not produce novelty, differentiation, or stabilised structure.

- Linear persistence ($\beta\mathscr{K}$) — A self-generating field must also retain continuity across iterations. The linear term represents this persistence: it carries forward the current stabilisation, preserving coherence between successive states. It embodies the memory or inertia of the field, ensuring that change is not chaotic but accumulative.

- Invariant baseline (γI) — Finally, every iteration presupposes a background of identity, the logical or physical "space" within which transformation occurs. The identity term supplies this invariant reference, the conserved quantity or normalisation against which all evolution is measured.

Taken together, these three components form the minimal closed algebra capable of describing the evolution of a self-referential generative process.

The map is quadratic, rather than of higher order, because self-interaction (action of the operator on itself) is the simplest nonlinearity sufficient to produce differentiation and feedback; and it is tensorial, because the field's evolution involves coupling between multiple interdependent channels — stability and fidelity, internal and external, real and imaginary components.

In this sense, the quadratic tensor recursion is the parsimonious canonical form of VGF dynamics. It stands to generative systems as the Schrödinger equation stands to quantum mechanics or the logistic map to population dynamics: the simplest expression that preserves reflexivity, continuity, and invariance while remaining closed under its own action.

All higher-order formulations or domain-specific equations — whether in physics, biology, or cognition — can be viewed as elaborations or particular instantiations of this same universal recursive grammar.

Thus,

$$g(\mathscr{K}) = \alpha\mathscr{K}^2 + \beta\mathscr{K} + \gamma I$$

is not an arbitrary choice, but the natural and minimal mathematical structure for describing a field that iterates upon itself. It is the algebraic skeleton of the VGF: a single form capable of generating both stability and transformation, coherence and decoherence, the persistent self-organisation of the field through its own infinite iteration.

Technical Note: Closure and Idempotence

From a formal standpoint, the quadratic tensor recursion

$$g(\mathscr{K}) = \alpha\mathscr{K}^2 + \beta\mathscr{K} + \gamma I$$

arises as the minimal polynomial closure of a self-referential operator.

Any self-acting process $\mathscr{K} \mapsto f(\mathscr{K})$ that remains stable under its own iteration:

$$\mathscr{K}_{n+1} = f(\mathscr{K}_n)$$

must admit fixed-point or idempotent solutions E satisfying:

$$E = f(E) = E^2.$$

The smallest algebra of endomorphisms closed under both linear combination and self-composition therefore contains precisely the powers $\{I, \mathscr{K}, \mathscr{K}^2\}$.

Higher-order terms $\mathscr{K}^n$ for $n > 2$ add no new structural degrees of freedom, since any iterative hierarchy of self-maps reduces, under idempotent completion (Karoubi envelope), to compositions of these three.

Hence a general second-degree polynomial is the canonical closure of reflexive iteration.

In categorical language, this corresponds to the Karoubi-complete envelope of a self-functor: the construction that adds all stable idempotents and closes the category under its own composition.

The coefficients α, β, γ then parameterise how much of the next morphism is drawn from the quadratic self-interaction, from linear persistence, or from the identity inclusion.

Because this polynomial is both algebraically closed and dynamically sufficient to generate fixed points, attractors, and bifurcations, it serves as the universal normal form for VGF evolution.

Stability-Fidelty

From the primitive act of iteration, the VGF (Vast Generative Field) is the closure of everything that can be produced by the generative rule under infinite composition. Formally, if $\mathcal{I}$ denotes the primitive iteration operator, then the VGF is the space of structures invariant under its repeated application. The essential idea is: whatever stabilises under unbounded iteration becomes a structure; whatever does not stabilise is transitory and does not form part of the persistent world.

The Stability-Fidelity Law

The framework therefore requires a clear sense of what "stabilisation" means. This is where the notion of closure enters. A closure is any structure that becomes idempotent under the action of iteration: applying the generative process again does not change it. Closures are the fixed points of the IIP — they are the first truly durable forms.

Mathematically, a closure E satisfies $\mathcal{I}(E) = E$. Conceptually, a closure is a kind of "environmental shell": a stabilised pattern that constrains further generativity and provides a background for the next layer of structure to arise. In the full VGF, closures form hierarchically: each new closure becomes the environment in which subsequent closures stabilise.

A central organising principle of the framework — one that recurs from physics to biology to cognition — is the Stability–Fidelity Law. This law expresses a universal trade-off: systems that achieve stability do so by sacrificing fine-grained fidelity to the underlying generativity. Stability arises through redundancy, repetition, and coarse-graining, whereas fidelity requires sensitivity to the full complexity of the generative process. The two cannot be maximised simultaneously.

The Stability-Fidelity Law that is central to the IIP-VGF framework can be expressed in a completely generic way as as a monotone trade-off under a coarse-graining map.

Let $\mathcal{G}_\lambda$ be a coarse-graining (or "environmental averaging") map at scale λ (classically: a Markov kernel / lumping operator; quantumly: a CPTP channel). Measure "fidelity to the fine/coherent description" by how much distinguishability survives, and "stability" by how contractive the map is.

A very generic, succinct form is:

$$\kappa(\mathcal{G}_\lambda) \ := \sup_{\rho \neq \sigma} \frac{D\big(\mathcal{G}_\lambda(\rho), \mathcal{G}_\lambda(\sigma)\big)}{D(\rho, \sigma)} \ \leq 1 \qquad \text{(data processing)}$$

where D is any divergence/metric that obeys data processing (*e.g.* relative entropy, trace distance, *etc.*).

Then define (one convenient choice)

$$\textbf{Fidelity}(\lambda) := \kappa(\mathcal{G}_\lambda), \qquad \textbf{Stability}(\lambda) := 1 - \kappa(\mathcal{G}_\lambda).$$

The Stability–Fidelity Law in coarse/fine-graining form becomes:

$$\lambda \uparrow \ \Rightarrow \ \kappa(\mathcal{G}_\lambda) \downarrow \quad \text{so} \quad \textbf{Stability}(\lambda) \uparrow, \ \textbf{Fidelity}(\lambda) \downarrow .$$

Interpretation:

Stronger coarse-graining (more redundancy/averaging) makes macrostates less sensitive to micro-perturbations (more stable) precisely because it suppresses fine-grained distinctions (lower fidelity to the underlying coherence).

We could also write this as:

$$I_{\text{ret}}(\lambda) := I(\text{micro}; \text{macro}_\lambda) \downarrow \iff S_{\text{rob}}(\lambda) \uparrow ,$$

i.e. retained micro-information decreases as robustness increases under coarse-graining.

Quantum Channel Application

In quantum physics, the Stability–Fidelity Law follows from CPTP contractivity: any channel that stabilises states by coarse-graining must reduce trace-distance distinguishability, gaining robustness precisely by losing fidelity to underlying quantum coherence.

The Stability–Fidelity Law in Quantum Channel Form)

Let

$$\Phi : \mathscr{B}(\mathscr{H}) \to \mathscr{B}(\mathscr{H})$$

be a quantum channel (completely positive, trace-preserving), representing environmental coupling, decoherence, or coarse-graining at some scale.

Define quantum distinguishability by the trace distance

$$D(\rho, \sigma) := \frac{1}{2}\|\rho - \sigma\|_1 .$$

By the data-processing (contractivity) theorem, for all density operators ρ, σ,

$$D(\Phi(\rho), \Phi(\sigma)) \ \leq \ D(\rho, \sigma).$$

Define the fidelity-retention coefficient

$$\kappa(\Phi) \ := \ \sup_{\rho \neq \sigma} \frac{D(\Phi(\rho), \Phi(\sigma))}{D(\rho, \sigma)} \ \in \ [0,1] .$$

Then the Stability–Fidelity relation may be written as

$$\textbf{Fidelity}(\Phi) := \kappa(\Phi), \qquad \textbf{Stability}(\Phi) := 1 - \kappa(\Phi) .$$

Stronger decoherence or coarse-graining corresponds to channels with smaller $\kappa(\Phi)$: they suppress fine-grained quantum distinctions (lower fidelity) while producing outputs that are increasingly insensitive to microscopic variation (higher stability).

Canonical decoherence example (pointer-basis attraction)

Let Δ be the dephasing channel in a pointer basis

$$\{|i\rangle\} : \Delta(\rho) := \sum_i |i\rangle\langle i| \rho |i\rangle\langle i| .$$

- Off-diagonal coherence is eliminated.
- Diagonal states form a stable fixed-point algebra.
- $\kappa(\Delta) < 1$, expressing loss of fidelity to coherent phase relations.

A continuous family,

$$\Phi_p(\rho) = (1 - p)\rho + p\,\Delta(\rho), \qquad p \in [0,1],$$

makes the trade-off explicit:

$$p \uparrow \Rightarrow \kappa(\Phi_p) \downarrow \Rightarrow \textbf{Stability} \uparrow , \textbf{Fidelity} \downarrow .$$

The First Physical Fidelity Cascade — from Spacetime to Energy to Particles

The IIP-VGF (**Infinite Iteration Principle - Vast Generative Field**) framework is a mathematical and scientific approach based on a single postulate or axiom - that infinite iteration is the fundamental primitive of nature.

In the IIP–VGF framework, nothing begins with ready-made objects, fields, or substances. What exists first as nature is **iteration itself** — ongoing generative activity — and the world as we know it arises only where that activity stabilises into persistent patterns.

A stabilised pattern is what the framework calls a **closure**: something that holds together across repetition strongly enough to behave as a recognisable structure.

The Vast Generative Field (VGF) is the field of closure or closures of infinite iteration under its own activity: an inexhaustible **generative coherence** within which all stabilised forms arise as **nested decohered shells**, none of which fully capture or contain the field itself.

Time arises as a closure of the IIP itself, before the full emergence of the VGF. It is the stabilisation of iteration into an ordered sequence of partially decohered states. It is not an external parameter but an emergent constraint on generative coherence.

Time, in the IIP–VGF framework, is the weakest form of closure: the stabilisation of iteration into ordered succession, arising through minimal **decoherence** of generative **coherence**.

In physics then, the first robust stabilisation in the VGF is what physics calls **spacetime** — not as a container into which things are later placed, but as the earliest form of **coherent persistence**. It is the first level at which processes can remain sufficiently stable for anything like a "world" to begin unfolding.

However, prior to robust closure, the question of "how many" has no meaning; therefore, to call **spacetime** the "first" closure is a conditional statement made only within an already-stabilised ordering. This already-stabilised ordering is the same one in which our own intelligence evolves.

Spacetime is the first **Fidelity shell** in continuing evolution by decoherence: a stabilised domain that arises through the loss of direct fidelity to the underlying generative coherence from which it emerges.

Each subsequent Fidelity shell stabilises only by further sacrificing fidelity to deeper coherence, trading generative exactness for robustness, persistence, and objectivity. Once such a spacetime-like coherence exists, new stabilisations become possible within it.

The next major one is what physics calls **energy**. In the IIP–VGF reading, energy is not a substance, nor a mysterious metaphysical entity, but a pattern of invariance: something that remains conserved even while processes change their form.

It expresses the fact that, inside the spacetime shell, certain aspects of dynamical behaviour can recur reliably and with fidelity. Energy is therefore a second-order closure — a stabilisation not of "things," but of how transformations themselves remain coherent across time.

From here, still more refined closures can arise. When these conserved dynamical structures become localised and repeatable, they stabilise as particles and fields. These are not independent things "in space", but local **coherence-patterns** within the broader energetic field — stable enough to appear redundantly again and again, to interact, and to participate in more complex structures.

They are, in effect, **localised closures of energy**, just as **energy is a closure of dynamical invariance**, and **spacetime is the closure that makes stable persistence in time, possible in the first place**.

Seen in this light, the familiar hierarchy of physics — spacetime, energy, particles — is not a list of different ontological ingredients, but the result of a **cascade of stabilisations**, each one made possible by the one before it.

Spacetime makes persistence possible; persistence allows invariant structure; invariant structure allows localised, repeatable entities. From those entities, chemistry, biology, ecosystems, nervous systems, and finally even intelligence and culture emerge as still higher-order closures built on top of the same generative pattern.

So the deep claim of the IIP–VGF framework is therefore that quantum evolution, cosmic evolution, biological evolution, and the evolution of intelligence are not separate processes governed by unrelated principles. Rather, they are successive manifestations of one underlying logic: the progressive **stabilisation of coherence through iteration**, **redundancy**, and **nested closure** — the world continually building itself layer upon layer, each layer both supported by and transforming the one beneath it.

The Cascade: Spacetime → Energy → Particles

In the IIP–VGF framework, persistence does not arise from pre-given entities, but from closures of iterative activity that stabilise into fidelity-preserving structures. The first such structures form a cascade of nested Fidelity shells, beginning with spacetime, continuing through conserved dynamical invariants (energy), and culminating in localised excitation-closures (particles and fields).

Definition (Spacetime Fidelity Shell)

Let $\mathcal{S} = (\mathcal{S}, \otimes, I)$ be a symmetric monoidal category equipped with a monoid object T representing time evolution.

A spacetime Fidelity shell is an object $E \in \mathcal{S}$ together with

1. an idempotent endomorphism $e : E \to E$ satisfying $e^2 = e$, and

2. an isomorphism $\varphi : E \xrightarrow{\cong} \mathbb{I}(E)$ making E a fixed point of the iterative functor $\mathbb{I}$, such that the compatibility condition

$$\mathbb{I}(e) \circ \varphi = \varphi \circ e$$

holds.

The full subcategory generated by E under $\otimes$ and admissible morphisms is denoted $\mathcal{S}_E$ and is called the spacetime shell sector.

Remark

The idempotent e expresses self-stability under refinement, while φ expresses stability of E under global iteration. This is the minimal envelope within which persistence of structure is possible.

Lemma (Persistence Enables Invariant Functionals)

Let E be a spacetime Fidelity shell and let $\mathcal{S}_E$ be its associated sector. Then there exists a strong monoidal functor

$$\mathcal{H} : \mathcal{S}_E \to (\mathbf{Ab}, \oplus, 0)$$

such that

$$\mathcal{H}(X \otimes Y) \cong \mathcal{H}(X) \oplus \mathcal{H}(Y), \qquad \mathcal{H}(I) = 0,$$

and for every admissible temporal evolution $f : X \to X$,

$$\mathcal{H}(f) = \mathrm{id}_{\mathcal{H}(X)}.$$

Proof (Sketch)

Because E is stable under iteration, its morphisms form a fidelity-preserving subcategory closed under composition and tensoring. By standard monoidal reconstruction, such sectors admit additive invariants that respect both tensor composition and temporal symmetry; these assemble into a strong monoidal functor $\mathcal{H}$.

Definition (Energy Closure)

An energy closure in $\mathcal{S}_E$ is a pair (X, h) with $X \in \mathcal{S}_E$ and $h \in \mathcal{H}(X)$ such that for every temporal morphism $f : X \to X$,

$$\mathcal{H}(f)(h) = h.$$

The value h is called the energy invariant of X.

Remark

Energy is here understood as a closure of dynamical invariance: an idempotent selection of those processes whose behaviour preserves a conserved quantity across iteration.

Lemma (Localisation of Invariants)

Let (X, h) be an energy closure and suppose X factorises as

$$X \cong \bigotimes_{i \in \Lambda} X_i.$$

Then $\mathscr{H}(X)$ decomposes as

$$\mathscr{H}(X) \cong \bigoplus_{i \in \Lambda} \mathscr{H}(X_i),$$

and there exist idempotent projections

$$p_i : \mathscr{H}(X) \to \mathscr{H}(X), \qquad p_i^2 = p_i,$$

selecting invariant local energy sectors.

Proof (Sketch).

Monoidality of $\mathscr{H}$ ensures additivity across tensor factors. The universal property of coproducts yields canonical projections, which are idempotent because invariance is preserved under iteration.

Definition (Particle / Field Closure)

A particle (or field excitation) closure in $\mathcal{S}_E$ is a triple (X_i, π_i, p_i) consisting of

1. an object $X_i \subseteq X$,

2. an idempotent endomorphism $\pi_i : X \to X$ with $\pi_i^2 = \pi_i$, and

3. a corresponding idempotent sector map $p_i : \mathscr{H}(X) \to \mathscr{H}(X)$

such that

$\mathscr{H}(\pi_i) = p_i$ and X_i is stable under iteration,

$\mathbb{I}(X_i) \cong X_i$.

Corollary (Nested Fidelity Cascade)

From a spacetime Fidelity shell E arise successively:

1. Energy closures as conserved monoidal invariants in $\mathcal{S}_E$;

2. Particle / field closures as localised idempotent sectors of those invariants.

Each closure forms a platform on which further higher-order closures (chemical, biological, cognitive) may stabilise as subsequent Fidelity shells.

Interpretation

Spacetime provides the first persistence-enabling closure; energy introduces conserved dynamical structure within it; particles and fields arise as robust, localised excitation-closures. The familiar physical hierarchy thus appears as a sequence of nested idempotent stabilisations generated by the same iterative principle.

Dimension as an Attractor in the IIP–VGF Framework

Introduction

In the IIP–VGF framework, spatial dimension is not a primitive parameter but an emergent attractor under coarse-graining. Prior to spacetime, the generative field possesses no defined dimensionality, only a high-dimensional relational state space subject to iteration.

As iterative dynamics are progressively coarse-grained, unstable microscopic correlations wash out while redundant structures stabilise, producing effective locality and metric behaviour.

Dimensionality becomes definable only at this stage, through the scaling laws of correlations and invariants. Renormalisation-group flow then selects stable fixed points, and three spatial dimensions emerge as a dynamically robust attractor: sufficiently rich to support locality, conservation, and complex structure, yet stable under further iteration and decoherence.

Spacetime dimensionality is therefore not assumed by the framework, but selected by the same stability–fidelity principle that governs all higher-order evolution within the VGF.

1. Starting point: pre-geometric generative structure

We begin prior to spacetime, with:

- an iterative generative field governed by the Infinite Iteration Principle (IIP),

- a very high-dimensional (possibly infinite-dimensional) state space $\mathscr{K}$,
- no distinguished notion of locality, metric, or spatial direction.

Formally, we can think of $\mathscr{K}$ as:

- a space of relational degrees of freedom,
- equipped with an iteration operator $\mathbb{I} : \mathscr{K} \to \mathscr{K}$,
- but without a preferred decomposition into spatial axes.

At this level:

- "dimension" is not defined,
- "space" does not yet exist,
- only iterative correlation structure exists.

2. Coarse-graining as loss of microscopic fidelity

Define a coarse-graining map

$$\mathscr{C}_\lambda : \mathscr{K} \to \mathscr{K}_\lambda$$

parameterised by a scale λ, where:

- increasing λ corresponds to discarding fine-grained relational detail,
- $\mathscr{K}_\lambda$ retains only stable, redundant correlations.

This is exactly the Stability–Fidelity trade-off in formal language:

- fidelity to microscopic structure is reduced,

- stability of macroscopic patterns is increased.

Notably:

- $\mathscr{C}_\lambda$ does not presuppose space,
- it is defined purely in terms of correlation redundancy under iteration.

3. Effective dynamics under renormalisation

Define an effective iteration operator at scale λ:

$$\square_\lambda := \mathscr{C}_\lambda \circ \square \circ \mathscr{C}_\lambda^{-1}$$

As λ increases, many microscopic degrees of freedom wash out, and the system flows toward effective descriptions that are invariant under further coarse-graining.

This defines a renormalisation-group flow on the space of effective theories:

$$\lambda \mapsto (\mathscr{K}_\lambda, \square_\lambda)$$

4. Emergence of locality and metric structure

At certain scales, the RG flow admits fixed-point structures where:

- correlations factorise approximately,
- interactions become short-range,
- and the system admits a notion of neighbourhood.

Formally, this corresponds to the emergence of a tensor-factorisation structure

$$\mathscr{K}_\lambda \approx \bigotimes_{i \in \Lambda} \mathscr{K}_{\lambda,i}$$

where Λ is an indexing set of effectively independent subsystems.

This is the birth of locality.

Once locality exists, one can define:

- adjacency relations on Λ,
- effective distances from correlation decay rates,
- and hence an emergent metric geometry.

Only here does it become meaningful to talk about dimension.

5. Dimension as an RG fixed point

Define an effective spatial dimension d via scaling behaviour, *e.g.*:

- correlation functions:

$$\langle \phi(x)\phi(y)\rangle \sim |x-y|^{-(d-2)}$$

- density of states:

$$N(R) \sim R^d$$

- or spectral dimension from diffusion:

$$P(t) \sim t^{-d/2}$$

These definitions do not assume a background space; they arise from scaling behaviour of correlations.

Now the key point:

Under RG flow, only certain dimensionalities are stable fixed points.

Formally:

- Let $\mathscr{D}_\lambda$ be the effective dimension at scale λ.
- The RG flow defines

$$\frac{d\mathscr{D}}{d \log \lambda} = \beta(\mathscr{D})$$

- Stable dimensions satisfy

$$\beta(\mathscr{D}^*) = 0, \quad \beta'(\mathscr{D}^*) < 0$$

Empirically and theoretically, d=3 is such a stable attractor:

- interaction laws are well-behaved,
- long-range structure survives,
- bound states exist,
- topological richness is maximal without instability.

Other dimensions correspond to:

- unstable fixed points,
- or runaway flows that fail to support persistent structure.

6. Interpretation in IIP–VGF terms

In IIP–VGF language:

- dimension is not chosen,
- it is selected by iterative stability.

Spacetime stabilises as the first Fidelity shell when the RG flow reaches a regime where:

- locality holds,
- geometry becomes robust,
- and dimensionality locks into a stable value.

Thus:

Dimension is an attractor of the iterative–decohering dynamics of the VGF.

3+1 spacetime is not fundamental — it is what survives coarse-graining.

Decoherence

Quantum decoherence can be read as an iteration–coarse-graining flow: the system repeatedly interacts with an environment and then the environment is traced out, yielding a quantum channel

$$\mathcal{E}(\rho) = \mathrm{Tr}_E[U(\rho \otimes \rho_E)U^\dagger].$$

Under iteration $\mathcal{E}^n$, phase information is progressively dispersed into environmental degrees of freedom, and the dynamics approaches an idempotent "closure"

$$\Pi(\rho) = \lim_{n\to\infty} \mathcal{E}^n(\rho)$$

projecting states onto the stable sector.

The resulting fixed structures — equivalently, the fixed-point algebra

$$\mathrm{Fix}(\mathcal{E}*)$$

— define the pointer observables and pointer projectors, *i.e.* the classical distinctions that survive repeated monitoring.

In the IIP–VGF view, this is the local, quantum-scale instance of the same mechanism by which spacetime geometry emerges: coarse-graining drives the generative field toward stable fixed-point structures, first selecting robust classical alternatives (pointer bases), and at larger scales selecting robust locality and scaling behaviour (effective geometry and dimensionality).

Both "preferred basis" and "preferred dimension" are therefore attractor properties of iterative dynamics under fidelity loss: what persists is what is stable under repeated interaction, redundancy, and coarse-graining.

In the Lindblad formulation of open quantum dynamics, decoherence is a semigroup

$$\mathscr{E}_t = e^{t\mathscr{L}}$$

acting on states, with dual action

$$\mathscr{E}_t^* = e^{t\mathscr{L}^*}$$

acting on observables. The observables that survive environmental monitoring form the fixed-point algebra

$$\mathrm{Fix}(\mathscr{E}^*) = \ker(\mathscr{L}^*),$$

often expressible as a commutant of the noise operators,

$$\{L_\alpha, L_\alpha^\dagger, H\}'.$$

Pointer projectors are then precisely the idempotents in this algebra,

$$\Pi_i^2 = \Pi_i,$$

yielding the stable classical alternatives. The long-time effect of decoherence is an idempotent "closure" map $\mathscr{P}$ (a conditional expectation) projecting $\mathscr{B}(\mathscr{H})$ onto $\mathrm{Fix}(\mathscr{E}^*)$, with $\mathscr{P}^2 = \mathscr{P}$.

Dually, ρ is projected onto its block-diagonal component

$$\sum_i \Pi_i \rho \Pi_i.$$

In IIP–VGF terms, this is the local quantum prototype of the same stability mechanism that later selects effective geometry: coarse-graining drives the theory toward fixed-point structures, first as a preferred classical subalgebra (pointer basis) and, at larger scales, as a preferred locality and scaling regime (spacetime dimension).

In both cases, "what survives" is precisely the idempotent sector selected by the flow.

The operator-algebraic structure of decoherence has a direct geometric analogue under coarse-graining. In the quantum case, the Heisenberg-picture semigroup

$$\mathscr{E}_t^* = e^{t\mathscr{L}^*}$$

selects a fixed-point algebra

$$\mathrm{Fix}(\mathscr{E}^*) = \ker(\mathscr{L}^*);$$

an idempotent conditional expectation projects general observables onto this stable subalgebra, and the pointer projectors are precisely the idempotents within it.

Geometric emergence can be formulated in the same way: coarse-graining defines a scale semigroup

$$R_\ell^* = e^{\ell\mathscr{G}}$$

acting on relational observables, and the "survivors" form a locality fixed-point algebra

$$\mathrm{Fix}(R^*) = \ker(\mathscr{G}).$$

The associated idempotent projection $\mathscr{Q}$ discards non-universal microstructure while retaining scale-invariant features. Dimension then appears not as a prior axiom but as a fixed-point exponent: at a stable geometric fixed point, volume growth and correlation scaling take the form

$$N(R) \sim R^d (or P(t) \sim t^{-d_s/2}),$$

so d is the emergent universality-class parameter of the attractor selected by the RG flow.

Thus pointer bases and spatial dimensionality are instances of the same IIP–VGF mechanism: iterative dynamics under fidelity loss selects an idempotent stable sector — first as a preferred classical subalgebra, and at larger scales as a preferred locality structure with a definite dimension.

In a VGF-native renormalisation reading, coarse-graining is an idempotent projection Π_ℓ acting on the relational state $\mathscr{K}$, and the effective dynamics at scale ℓ is obtained by conjugating the generative recursion

$$g(\mathscr{K}) = \alpha\mathscr{K}^2 + \beta\mathscr{K} + \gamma I$$

through Π_ℓ, yielding a scale-dependent recursion g_ℓ of the same functional family but with running couplings

$$(\alpha_\ell, \beta_\ell, \gamma_\ell).$$

The resulting flow of effective couplings defines a renormalisation-group map whose fixed points correspond to stable Fidelity shells.

Locality appears when the fixed point supports a robust factorisation of degrees of freedom into weakly coupled substructures, and dimensionality is then not a prior assumption but a fixed-point exponent: at a geometric universality class, neighbourhood growth obeys $N(R) \sim R^d$, so the spatial dimension d is part of the eigen-structure of the fixed point.

In this sense, "three dimensions" is what survives: it labels the stable geometric attractor of the VGF coarse-graining flow that can sustain long-lived nested closures, just as pointer bases label the stable attractors of decoherence at the quantum scale.

Space as an Idempotent Decomposition

If the coarse-graining flow R_ℓ^* admits a nontrivial fixed-point subalgebra

$$\mathscr{A}_{\text{loc}} = \ker(\mathscr{G})$$

that is approximately commutative at macroscopic scales, then $\mathscr{A}_{\text{loc}}$ determines an emergent topological space X and its stable regions *via* idempotents $p \in \mathscr{A}_{\text{loc}}$.

A resolution-limited set of minimal idempotents $\{p_x\}$ forms a spatial partition, and residual couplings between these idempotents induce adjacency and metric structure. Spatial dimension is then recovered as the scaling exponent $N(R) \sim R^d$ of neighbourhood growth in the induced adjacency geometry.

Decoherence selects a preferred classical subalgebra

$$\ker(\mathscr{L}^*)$$

whose idempotents are the pointer projectors. The geometric transition can be formulated in exactly the same operator-algebraic pattern: coarse-graining defines a scale flow

$$R_\ell^* = e^{\ell \mathscr{G}}$$

on relational observables, and the locality-stable survivors form a fixed-point algebra

$$\mathscr{A}_{\mathrm{loc}} = \ker(\mathscr{G}).$$

An idempotent projection $\mathscr{Q}$ maps arbitrary observables onto $\mathscr{A}_{\mathrm{loc}}$, just as the conditional expectation maps onto pointer observables. When $\mathscr{A}_{\mathrm{loc}}$ becomes approximately commutative, it is isomorphic to an algebra of functions on an emergent space X; its idempotents then define stable regions, and near-fixed residual couplings between these regional idempotents generate adjacency, metric, and dimensional scaling.

In this sense, "space" is the pointer decomposition of the locality algebra, and "dimension" is the universality-class exponent of the fixed point selected by iterative coarse-graining within the VGF.

The Birth of Spacetime

At the earliest geometric stabilisations, the locality-stable sector selected by coarse-graining need not be commutative; rather, the VGF yields a noncommutative locality algebra

$$\mathscr{A}_{\mathrm{loc}} = \ker(\mathscr{G})$$

whose idempotents represent stable "regions" that are not yet jointly sharp.

Classical space arises only when, under increasing redundancy and fidelity, a large subalgebra becomes approximately commutative, with commutators suppressed below a scale-dependent threshold $\varepsilon(\ell)$.

In that almost-commutative regime, the centre $Z(\mathscr{A}_{\mathrm{loc}})$ provides a canonical proto-classical base space, while metric and dimensional structure are extracted from spectral data — *e.g.* from a Dirac-like operator D_ℓ defining distance *via* commutator bounds and dimension via spectral scaling.

The familiar 3D spatial manifold is thus interpreted as the stable attractor reached when the noncommutative locality algebra flows, under VGF coarse-graining, into an almost-commutative fixed point whose spectral dimension locks to three.

Why commutativity should not be assumed at the first geometric closure

In the previous step we used the identification

$$\mathscr{A}_{\mathrm{loc}} \cong C(X)$$

to recover an ordinary space X. That requires $\mathscr{A}_{\text{loc}}$ to be commutative. But in the IIP–VGF framework, the earliest "geometric" stabilisations are still pre-classical: they are closures of correlation structure under coarse-graining, not yet a fully classical manifold.

Therefore the more faithful statement is:

The first locality-stable sector $\mathscr{A}_{\text{loc}}$ is generally noncommutative, and classical space arises only when $\mathscr{A}_{\text{loc}}$ becomes approximately commutative under increasing fidelity (redundancy).

This is exactly parallel to quantum decoherence: classical alternatives appear when interference terms are suppressed, not because the underlying theory becomes literally classical.

The noncommutative locality algebra and its "classical limit"

Let

$$\mathscr{A}_{\text{loc}} \subseteq \mathscr{O}$$

be the fixed-point algebra of the coarse-graining flow:

$$\mathscr{A}_{\text{loc}} := \ker(\mathscr{G}) = \{F : \ R_{\ell}^{*}(F) = F\}.$$

Now allow $\mathscr{A}_{\text{loc}}$ to be a (generally) noncommutative C^{*}-algebra.

Define its centre:

$$Z(\mathscr{A}_{\text{loc}}) := \{A \in \mathscr{A}_{\text{loc}} : \ [A, B] = 0 \ \forall B \in \mathscr{A}_{\text{loc}}\}.$$

- If $\mathscr{A}_{\text{loc}}$ is fully commutative, then $Z(\mathscr{A}_{\text{loc}}) = \mathscr{A}_{\text{loc}}$, and we recover an ordinary space.
- If not, then the centre gives the "most classical" part available at that stage.

Thus one can already define a proto-classical space as the spectrum of the centre:

$$X_{\text{cent}} := \operatorname{Spec}\big(Z(\mathscr{A}_{\text{loc}})\big).$$

This yields a canonical decomposition:

$$\mathscr{A}_{\text{loc}} \approx \int_{x \in X_{\text{cent}}}^{\oplus} \mathscr{A}_x \, d\mu(x),$$

a direct integral of noncommutative fibres $\mathscr{A}_x$ over a classical base.

Intuitively: a classical "base space" begins to appear, but each "point" still contains internal noncommutative structure.

Almost-commutativity as a fidelity condition (the key VGF move)

To connect this to Fidelity shells, we introduce an explicit fidelity threshold $\varepsilon(\ell)$ at scale ℓ. Define a commutator norm:

$$\Delta_\ell(A, B) := \Big\| [A, B] \Big\|_\ell,$$

where $\| \cdot \|_\ell$ is an effective norm appropriate to scale ℓ (*e.g.* induced by the coarse-grained state / measure).

Definition (Almost-Commutative Regime).

We say $\mathscr{A}_{\text{loc}}$ is ε-commutative at scale ℓ if there exists a large subalgebra

$$\mathscr{A}_{\text{cl}}^{(\ell)} \subseteq \mathscr{A}_{\text{loc}}$$

such that

$$\|[A,B]\|_\ell \le \varepsilon(\ell) \quad \text{for all } A,B \in \mathscr{A}_{\mathrm{cl}}^{(\ell)}.$$

Interpretation:

- $\varepsilon(\ell)$ is the residual nonclassicality allowed at that shell.
- "Classical spacetime" corresponds to $\varepsilon(\ell) \to 0$ (or becoming negligibly small) over the degrees of freedom relevant to macroscopic structure.

This is the direct geometric analogue of decoherence suppressing off-diagonals below experimental detectability.

Noncommutative regions: idempotents survive, but need not commute

In the commutative case, region projectors p_x commute and give ordinary partitions. In the noncommutative case, we still have idempotents / projections

$$p^2 = p, \quad p^\dagger = p,$$

but they may fail to commute: $[p,q] \neq 0$. These are quantum regions: they represent stable sectors, but not jointly sharp classical regions.

This is exactly the same logic as quantum logic:

- commuting projections = simultaneously definite classical events,
- noncommuting projections = incompatible "events".

Thus pre-classical geometry has "regions" in the sense of stable idempotents, but their incompatibility reflects nonclassical structure.

Geometry from spectral data: Connes-style "spectral triple" (VGF version)

Noncommutative geometry packages "space" by replacing points/coordinates with algebraic and spectral data. The standard structure is a spectral triple $(\mathscr{A}, \mathscr{H}, D)$.

Definition (VGF Spectral Triple, heuristic form).

A VGF spectral geometry at scale ℓ is a triple

$$(\mathscr{A}_{\text{loc}}, \mathscr{H}_{\ell}, D_{\ell}),$$

where:

- $\mathscr{A}_{\text{loc}}$ acts on the Hilbert space $\mathscr{H}_{\ell}$ of effective degrees of freedom at scale ℓ,
- D_{ℓ} is a scale-dependent "Dirac-like" operator encoding the effective adjacency/metric structure.

Then the analogue of distance is defined by Connes' formula:

$$d_{\ell}(\varphi, \psi) = \sup\left\{ |\varphi(A) - \psi(A)| : A \in \mathscr{A}_{\text{loc}}, \ \|[D_{\ell}, A]\| \le 1 \right\},$$

whereφ, ψ are states on $\mathscr{A}_{\text{loc}}$.

Key point: distance is defined without assuming a manifold. It is defined from the algebra and spectral constraints.

As coarse-graining increases, D_{ℓ} flows toward a form that makes the commutators small on the effective classical subalgebra, yielding an ordinary metric space in the limit.

Dimension as spectral dimension in the noncommutative regime

In a pre-classical regime, "dimension" is best defined spectrally. If D_ℓ is Dirac-like, then its eigenvalue growth defines a dimension analogue (Weyl scaling). Alternatively, define a diffusion operator Δ_ℓ and extract the spectral dimension from heat kernel scaling:

$$\operatorname{Tr} e^{-t\Delta_\ell} \sim t^{-d_s(\ell)/2}.$$

Then:

- $d_s(\ell)$ can "run" with scale (fractal / pre-geometric regimes),
- and it locks onto a stable value when the locality fixed point is reached.

Thus 3D is the stable spectral / volume exponent of the attractor.

How classical 3D space appears (the key convergence)

Putting the above together:

1. Coarse-graining selects a stable noncommutative locality algebra

$$\mathscr{A}_{\text{loc}} = \ker(\mathscr{G}).$$

2. Increasing redundancy (fidelity shell refinement) drives the system into an almost-commutative regime:

$$\|[A, B]\|_\ell \leq \varepsilon(\ell) \quad \text{on } \mathscr{A}^{(\ell)}_{\text{cl}}.$$

3. In that regime, the spectrum of $Z(\mathscr{A}_{\text{loc}})$ (or of $\mathscr{A}^{(\ell)}_{\text{cl}}$) behaves like an ordinary space, and the spectral metric induced by D_ℓ converges to a classical metric.

4. Dimension, read off from spectral scaling, locks to $d = 3$ as an RG attractor.

So "space" is not born classical: it becomes classical as the locality algebra becomes (approximately) commutative.

In quantum mechanics:

- algebra first (noncommuting observables),
- classical outcomes appear when a commuting subalgebra stabilises.

In spacetime emergence:

- algebra first (noncommuting locality observables),
- classical space appears when an approximately commuting subalgebra stabilises.

In both cases:

- classicality = commutativity + redundancy,
- the actuality of nature = what survives iteration under loss of fidelity.

So the IIP-VGF framework is saying:

The same principle that produces classical measurement outcomes in quantum mechanics also produces classical space.

The Relation of Complex Phase Closure and Geometric Closure

Definition A (Complex Phase Closure)

A complex phase closure is a stabilised sector of the VGF in which reversible iteration admits an internal U(1) symmetry and a norm-preserving inner product. Formally, it consists of a $\mathbb{C}-$Hilbert space $(\mathscr{H}, \langle \cdot, \cdot \rangle)$ together with a one-parameter group of unitary transformations

$$U(t) = e^{-itH}, \qquad U(t)^{\dagger}U(t) = I,$$

such that

$$\langle U(t)\psi, U(t)\phi \rangle = \langle \psi, \phi \rangle \quad \text{and} \quad \|\psi\|^2 := \langle \psi, \psi \rangle \text{ is invariant.}$$

This closure is the minimal algebraic stabilisation that supports phase, interference, and unitary composition without introducing a preferred basis.

Definition B (Geometric / Spacetime Closure)

A geometric closure is a locality-stable fixed-point sector selected by coarse-graining. Let $R_{\ell}^{*} = e^{\ell \mathscr{G}}$ be a coarse-graining flow acting on relational observables. The locality algebra is the fixed-point algebra

$$\mathscr{A}_{\text{loc}} := \ker(\mathscr{G}) = \{A : \ R_{\ell}^{*}(A) = A \ \forall \ell \geq 0\}.$$

Classical space appears when $\mathscr{A}_{\text{loc}}$ contains a large approximately commutative subalgebra $\mathscr{A}_{\text{cl}} \subseteq \mathscr{A}_{\text{loc}}$ such that

$$\|[A, B]\| \leq \varepsilon \quad \forall A, B \in \mathscr{A}_{\text{cl}},$$

with ε below the fidelity threshold of the shell. In the commutative limit,

$$\mathscr{A}_{\text{cl}} \cong C(X),$$

so that idempotents $p^2 = p \in \mathscr{A}_{\text{cl}}$ define stable spatial regions.

Proposition (Interlocking Closures and Two Measures of What Survives)

When complex phase closure and geometric closure interlock—i.e. when local factorisation becomes meaningful (via emergent regions X) and local quantum dynamics is supported (via $\mathbb{C}$-Hilbert structure) — two invariant measures arise, each quantifying what survives under iterative dynamics with fidelity loss:

1. Born weights (internal measure).

Decoherence selects a pointer decomposition $\{\Pi_i\}$ (orthogonal idempotents) on $\mathscr{H}$,

$$\Pi_i^2 = \Pi_i, \quad \Pi_i\Pi_j = 0 \ (i \neq j), \quad \sum_i \Pi_i = I,$$

and the unique unitary-invariant, additive weight compatible with this idempotent structure is

$$w_i(\psi) = \langle \psi, \Pi_i \psi \rangle,$$

yielding $w_i = |c_i|^2$ in the pointer basis $\psi = \sum_i c_i |i\rangle$

2. Dimensional scaling (external measure).

Coarse-graining selects a stable locality structure with neighbourhood growth

$$N(R) \sim R^d,$$

where d is the emergent spatial dimension as the fixed-point scaling exponent of the locality closure.

Thus probability and dimensionality occupy analogous roles: both are invariant measures associated with stable idempotent/fixed-point structure, the former internal to phase-coherent Hilbert dynamics and the latter external to emergent geometry.

Corollary (Why ℂ and Spacetime Are Neither Reducible to the Other)

The complex numbers are not "derived from spacetime," nor is spacetime derived from complex numbers. Rather, ℂ and spacetime are distinct closures of the VGF — phase coherence and locality — whose interlocking enables classical reality. Born weights and spatial dimension are the corresponding measures of survival selected by these coupled stabilisations.

Measurement, Probability and What Survives

One of the motivations for the IIP–VGF framework is to place the emergence of "classical reality" — both in measurement and in spacetime — within a single generative principle. We will now clarify what that claim does and does not entail with respect to the quantum measurement problem and the Born rule.

Measurement as Stabilisation, Not Collapse

In standard quantum mechanics, measurement appears to require a special postulate: the collapse of the wavefunction. Decoherence and quantum Darwinism explain the physical origin of effective collapse, but they do not remove the collapse postulate from the standard axiomatic formulation of quantum mechanics.

In the IIP–VGF framework, no such fundamental collapse is introduced. Instead, measurement is understood as an instance of iterative stabilisation under loss of fidelity. When a quantum system repeatedly interacts with an environment, phase coherence between certain components of the state is progressively dispersed into environmental degrees of freedom. What remains are stable, effectively classical alternatives — pointer states — corresponding to idempotent sectors of the system's observable algebra.

From this perspective, a "measurement outcome" is not a primitive event, but a closure: a structure that survives repeated interaction and decoherence. Collapse is not a fundamental dynamical process, but the phenomenological appearance of an idempotent projection onto a stable sector. This is directly analogous to the way classical space appears when noncommutative locality stabilises into an approximately commutative algebra.

Probability as a Measure of Stability

The IIP–VGF framework therefore addresses one central aspect of the measurement problem: why definite outcomes occur at all. Outcomes exist because only certain distinctions are robust under iteration. However, the framework does not attempt to derive numerical probabilities from first principles alone. Instead, it constrains where probabilities must come from and what form they must take.

In this view, probability cannot be interpreted merely as subjective ignorance or as an external frequency assignment. Rather, it must reflect the degree of stabilisation and redundancy with which a given outcome is supported across system–environment interactions. Probability is thus tied to what survives, not to what is momentarily possible.

This already imposes strong structural requirements. Any admissible probability rule must be:

- invariant under changes of basis,
- compatible with composition and coarse-graining,
- and consistent with the emergence of objectivity through redundancy.

These requirements severely restrict the form that probability assignments can take.

The Born Rule: Constraint Without Derivation

The Born rule — assigning probabilities proportional to squared amplitudes — satisfies precisely these constraints. Its quadratic form reflects the geometry of Hilbert space: orthogonality, inner products, and norms determine both distinguishability and redundancy under decoherence. In this sense, the Born rule is naturally aligned with the stabilisation mechanism described by the IIP–VGF framework.

However, the framework does not claim to derive the Born rule in isolation. Rather, it shows why any viable probability rule must occupy the same structural role as the Born rule: a measure invariant under the transformations that preserve stable closure.

The Born rule does not appear in this framework as an arbitrary axiom appended to quantum mechanics, but as the unique measure compatible with the emergence of "classical reality". Once outcomes are understood as idempotent closures stabilised under iterative decoherence, probability must quantify the relative robustness of those closures.

The squared Hilbert-space norm is the only measure that is invariant under unitary transformation, additive over decohered alternatives, and correctly scaled with redundancy. In this sense, the Born rule occupies the same structural role in measurement theory that spatial dimensionality occupies in geometry: both are invariant measures associated with stable fixed points of the generative process

In the IIP–VGF framework, the appearance of squared amplitudes is not an additional probabilistic axiom but a consequence of the quadratic tensor recursor

$$g(\mathscr{K}) = \alpha\,\mathscr{K}^2 + \beta\,\mathscr{K} + \gamma I$$

that governs stabilisation: because classical structures arise through repeated self-composition and redundancy, their relative robustness is measured by the unique scalar invariant that compounds correctly under quadratic iteration.

A Shared Origin with Classical Space

The deeper claim of the framework is not that quantum measurement and spacetime emergence are identical phenomena, but that they are governed by the same generative logic. In measurement, iterative decoherence selects a preferred classical subalgebra and its idempotent projections; in spacetime emergence, coarse-graining selects a locality algebra and its spatial idempotents. In both cases, classical structure arises because non-surviving distinctions are suppressed, and what remains is what can be redundantly stabilised.

From this standpoint, probability and dimensionality play analogous roles. Both are invariant measures associated with stable fixed points of the generative process: probability measures the relative robustness of measurement outcomes, while dimension measures the scaling structure of emergent space.

Neither is fundamental in isolation; both are expressions of what survives under iteration.

Review

The IIP–VGF framework does not solve the measurement problem by adding new postulates or redefining probability. Instead, it reframes the problem by identifying a deeper common principle: "classical reality" emerges through stabilised closure under iterative dynamics with loss of fidelity. Measurement outcomes and classical spacetime are two manifestations of this principle at different levels. The Born rule is not derived outright, but is shown to be structurally natural — indeed, difficult to avoid — once probability is understood as a measure of stabilisation rather than of ignorance.

In this sense, the framework does not replace quantum mechanics, but clarifies why its most puzzling features take the form they do: they are the shadows cast, at different scales, by the same generative process determining what survives.

Measurement as Idempotent Survival

Measurement in quantum mechanics is often presented as a special process requiring additional postulates: collapse, projection, or the assignment of probabilities by fiat. In the IIP–VGF framework, no such additions are required. Measurement is understood as the endpoint of a more general stabilisation process: the selection of structures that survive repeated interaction under fidelity loss.

The defining feature of a measurement outcome is not that it occurs once, but that it can be recovered repeatedly. A measured value must be readable by many observers, at different times, through different interactions, without further degradation. This requirement introduces redundancy, and redundancy fundamentally changes the stabilisation problem. What survives must survive being composed with itself under coupling to an environment and coarse-graining.

This makes measurement intrinsically quadratic.

Let Φ denote the system–environment interaction, and let π represent the fidelity shell projection that retains only redundantly accessible information while discarding phase relations that cannot be stably copied. The effective stabilisation map is then not linear evolution alone, but the quadratic tensor recursor

$$\mathscr{R}(x) = \pi\big(\Phi(x)\,\Phi(x)\big),$$

which tests whether a candidate structure x remains invariant under repeated interaction and projection.

Stable measurement outcomes correspond to fixed points of this recursor:

$$\mathscr{R}(x) = x \quad \text{(up to shell equivalence).}$$

Such fixed points are generically idempotent:

$$x^2 = x.$$

Idempotency expresses the defining property of a classical outcome: once stabilised, further repetition adds nothing new. The outcome is self-consistent under its own reproduction.

In quantum measurement, these idempotent survivors appear as pointer projectors $\{\Pi_i\}$. They define mutually exclusive, redundantly recordable sectors of the system–environment state, and satisfy

$$\Pi_i^2 = \Pi_i, \qquad \Pi_i\Pi_j = 0\ (i \neq j).$$

Measurement does not select a result by collapse; it selects idempotent sectors by survivability.

Probability then enters naturally. When multiple idempotent survivors are possible, the question is not "which outcome is real?", but how strongly each outcome is stabilised under repeated interaction. The only measure compatible with quadratic self-composition is the squared norm:

$$p_i = \|\Pi_i \psi\|^2.$$

This quantity is not an arbitrary axiom, but the unique weight that is preserved under redundancy-building iteration. It measures the degree to which a candidate outcome survives being copied, dispersed, and re-encountered. The Born rule therefore expresses relative stability, not subjective ignorance.

Seen this way, measurement, decoherence, and probability are not separate puzzles. They are successive aspects of the same principle. Measurement outcomes are idempotent survivors; probabilities quantify their relative robustness; objectivity arises from redundancy. What becomes classical is precisely what can survive iteration with fidelity loss.

Measurement is therefore not where quantum mechanics breaks down. It is where the general logic of the VGF becomes visible: what survives under repeated interaction becomes real.

In this sense, **measurement is not the imposition of classicality on a quantum world, but the moment when survival itself becomes observable**.

Gravity

In the IIP-VGF framework then, gravity is not a force, nor a field in the first instance. It is not a primitive interaction between spacetime and matter.

Rather, gravity is the residual constraint imposed by the stabilised spacetime closure on **the persistence of structure** (other VGF closures). Here, "structure" refers to any stabilised pattern of energy, motion, or relational organisation that persists across iteration within spacetime.

Particles, fields, bound systems, and trajectories are all instances of such structure — not as primitive entities, but as **closures** that **maintain identity** by **remaining coherent under interaction**.

Gravity is therefore not a force acting on objects, but the constraint imposed on spacetime itself so that these persistent structures can continue to exist without disrupting the coherence of the spacetime fidelity shell.

It is necessary therefore to think go gravity not as coupled to **things**, but coupled to **persistence** (in time).

There are certain salient points to be made initially:

Why Gravity is Geometric

Gravity is what it costs, geometrically, for spacetime to continue to be a stable envelope in the presence of energy and matter. This is why gravity couples universally - it is not an interaction between things, but a constraint on everything that **persists**.

In the IIP-VGF framework spacetime emerges as a l**ocality-stable fixed point**.

As we have seen, dimensionality is an attractor. Geometry is a commutative limit of a locality algebra. Once that closure has occurred, the only way spacetime can respond to internal structure (energy, particles, fields) without breaking its own stability is by adjusting its geometric relations.

So gravity is not "matter pulling on matter" as such, but rather, the reconfiguration of locality required to keep spacetime itself coherent. This recovers the Einstein insight naturally, but explains why geometry, rather than a force, had to be the medium.

Why gravity is tied to energy, not charge or type

Other interactions (electromagnetic, weak, strong):

- depend on internal symmetries,
- distinguish particle types,
- act within spacetime.

Gravity, by contrast:

- couples to energy–momentum,
- cares only about persistence and motion,
- ignores internal quantum numbers.

In terms of the IIP–VGF framework energy is the next closure inside spacetime — a conserved invariant of transformation. So gravity is **the coupling between the spacetime closure and the energy closure nested within it**.

That is why everything with energy gravitates, and nothing escapes it.

Why gravity is weak (but never goes away)

From this perspective, gravity's weakness is no mystery. Gravity is not a primary stabilisation. It is a secondary corrective constraint. It only becomes noticeable when **large-scale coherence** is threatened. Gravity is what remains after all other stabilisations have done their work.

So it becomes negligible at particle scales, dominant at cosmological scales, **and impossible to screen**. We cannot simply turn off the requirement that spacetime remains coherent.

Why gravity resists quantisation

If gravity were just another field, then we should expect it to quantise like the others. But the IIP–VGF framework shows something different:

Gravity is a response of the spacetime closure, not a degree of freedom inside it.

Quantising gravity is therefore like trying to quantise dimensionality, locality itself, or the existence of a classical measurement outcome. This doesn't mean quantum gravity is impossible — but it does mean it cannot be a naive "graviton field on spacetime."

In contrast, the IIP-VGF framework predicts that any viable quantum gravity must describe:

how the spacetime closure itself fluctuates, decoheres, and stabilises — not how a force propagates within a fixed background.

*

So in the IIP–VGF framework gravity is the **manifestation of spacetime's self-stabilisation in the presence of nested closures**. It is the geometric response required to preserve locality, persistence, and coherence when energy — understood as a conserved invariant of transformation — occupies and flows through the spacetime fidelity shell. Gravity is what spacetime does to remain spacetime - it is **what survives** when **spacetime must accommodate further structure**.

It is true to say "gravity is spacetime curvature", but the IIP-VGF framework explains:

- why curvature appears at all,
- why it couples universally,
- why it is weak,
- why it resists quantisation,
- and why it is inseparable from spacetime emergence.

Quite simply, gravity is part of the cascade:

Iteration → Spacetime closure → Energy closure → Gravity as coupling

Gravity is not a force acting within spacetime by the bending of space time, but rather, it is the way spacetime maintains its own stability when burdened with energy. It is the price paid for **persistence** once **locality has emerged**.

The Equivalence Principle

In the IIP–VGF framework, the equivalence principle expresses the identity between inertia and gravity as two aspects of the same stabilisation. Inertia describes how spacetime supports coherent persistence of structures, while gravity describes how spacetime must reconfigure that support in the presence of energy.

Because both refer to the same **closure condition**, they cannot differ, and free fall corresponds to motion along locally stabilised persistence paths of the spacetime fidelity shell.

The equivalence principle holds because gravity is not fundamentally something acting on motion — rather, it is the rule by which **motion remains coherent** in already existent spacetime.

We could say:

Inertia is how spacetime lets things move; gravity is how spacetime adapts when things exist.

This connection explains how the equivalence principle is not an unexplained axiom, but rather, is a direct consequence of spacetime being a stabilised closure, and gravity being a secondary response rather than a fundamental interaction.

The equivalence principle is what we would expect from the underlying thesis of the IIP-VGF framework, that the deepest principles of physics are not forces or particles, but conditions of **persistence under iteration**. It is about what survives in nature's fundamental principle of creation through iteration.

The Cosmological Constant

In the IIP–VGF framework, dark energy / the cosmological constant is not a mysterious substance, field, or vacuum fluid, but a statement about the global self-consistency of the spacetime closure itself. Specifically, dark energy is the residual baseline constraint required for spacetime to remain a stable fidelity shell once locality has emerged.

Why the cosmological constant looks so strange in standard physics

In standard formulations, the cosmological constant Λ:

- acts everywhere, uniformly,
- does not dilute with expansion,
- couples to spacetime itself rather than to matter,
- produces acceleration rather than attraction,
- and has an absurdly small but nonzero value.

This makes it anomalous if interpreted as:

- vacuum energy of quantum fields,
- a new force,
- or a dynamical component like matter or radiation.

Those interpretations struggle because they treat Λ as something inside spacetime.

The IIP–VGF framework: Λ is about spacetime, not contents

In the IIP-VGF framework spacetime is not given, it is a stabilised closure, and gravity is spacetime's self-adjustment to internal structure.

Once this is realised, a new category becomes available: There may be constraints required for spacetime to exist at all, independently of local matter.

Dark energy is exactly such a constraint. It is not energy in spacetime; it is **the cost of maintaining spacetime as a coherent envelope**.

Why dark energy is uniform and unavoidable

Spacetime, once stabilised, must satisfy global consistency conditions:

- locality must remain well-defined,
- causal structure must remain coherent,
- the fidelity shell must not collapse under its own coarse-graining.

These conditions are **global**, not local. So the associated term must:

- act everywhere,
- be isotropic and homogeneous,
- not depend on local particle content.

That is precisely the behaviour of a cosmological constant. In IIP-VGF terms:

Λ is the **zeroth-order stabilisation term of the spacetime closure**. Matter and radiation are perturbations within the shell; Λ is part of the shell's own equilibrium condition.

Why dark energy causes expansion rather than attraction

Despite an initial initially counterintuitive feel, what follows naturally is that:

Spacetime is stabilised through coarse-graining:

- degrees of freedom are suppressed,
- redundancy is amplified,
- locality emerges as an attractor.

Coarse-graining generically produces:

- effective repulsion at large scales,
- dilution of correlations,
- smoothing of structure.

So in the IIP-VGF picture accelerated expansion is not "spacetime being pushed apart," but rather, it is spacetime **relaxing toward its stable large-scale configuration**. Dark energy is the signature of that relaxation.

Why Λ is small but nonzero

If Λ were:

- zero → spacetime would be marginally stable, highly sensitive to perturbation,
- large → spacetime would never have supported long-lived structure.

A universe with galaxies, chemistry, and observers requires:

- spacetime to be almost self-sustaining,
- but not exactly scale-invariant.

So in IIP–VGF terms:

Λ measures how close spacetime is to a fixed point of the generative process. Small Λ = near-criticality.

This replaces fine-tuning with survivorship:

Only universes where spacetime closure is barely but sufficiently stable can host higher-order closures.

Why dark energy is not screened or localised

Λ is not an interaction. It is not mediated, it is not exchanged, and it is not carried by particles. Rather, it is a background constraint, like dimensionality.

We cannot screen dimensionality; we cannot shield locality; and we cannot turn off the requirement that spacetime remain coherent.

Thus dark energy:

- does not clump,
- does not form structures,
- does not dilute.

Relation to gravity and the equivalence principle

Matter tells spacetime how to curve locally. Dark energy tells spacetime how to exist globally.

In the same way that:

- gravity is spacetime responding to energy,
- inertia is spacetime supporting motion,

dark energy is spacetime maintaining its own baseline configuration.

So the hierarchy becomes:

$$\text{Spacetime closure} \;\Rightarrow\; \begin{cases} \text{Cosmological constant (baseline stability)} \\ \text{Gravity (response to energy)} \end{cases}$$

They are not separate mysteries; they are two levels of the same mechanism.

Summary

Thus we see that In the IIP–VGF framework, the cosmological constant does not represent vacuum energy or a new physical substance, but a global **stabilisation term** associated with the spacetime fidelity shell itself.

It expresses the residual constraint required for locality and persistence to remain coherent at the largest scales. Matter and radiation deform spacetime locally, producing gravity; the cosmological constant fixes spacetime's baseline configuration globally. Its small but nonzero value reflects near-critical stability:

Spacetime is sufficiently robust to support nested closures, yet not exactly scale-invariant.

We can say that:

Gravity is spacetime reacting to what exists. Dark energy is spacetime ensuring that existence remains possible. It is again about **survival**. Dark energy is the signature of spacetime **surviving its own coarse-graining**.

Thus, in the IIP-VGF framework we:

- dissolve the "vacuum catastrophe",
- explain uniformity without new fields,
- explain smallness without fine-tuning,
- align with emergence of spacetime,

Again, we see that the deepest constants of nature are not parameters of matter, but **conditions of persistence**.

Cosmic Acceleration Data in the IIP-VGF Framework

Cosmic Acceleration as a Stability Diagnostic of Spacetime

In the IIP–VGF framework, observations usually described as evidence for "cosmic acceleration" are reinterpreted as empirical probes of the global stability condition of the spacetime fidelity shell.

What the data constrain is not a new form of matter or an exotic force, but the large-scale self-consistency of spacetime once locality has emerged. In standard cosmology this consistency condition is parameterised by the cosmological constant or, more generally, by an equation-of-state parameter w(z). In VGF terms, these parameters quantify how close the spacetime closure lies to a fixed point of the generative process.

Type Ia supernovae measure luminosity distances and show that the late-time expansion history deviates from that of a purely matter-dominated universe. Baryon acoustic oscillations provide a standard ruler that maps the expansion rate and distance scales across cosmic time. The cosmic microwave background fixes the early-universe boundary conditions against which this entire history is calibrated.

Taken together, these datasets reconstruct the function H(z) and reveal that spacetime's large-scale behaviour is consistent with a nearly constant baseline term, corresponding to $w \approx -1$, with current data allowing only mild departures from exact constancy.

Within the IIP–VGF framework, this observational situation has a natural interpretation. A perfectly vanishing baseline would correspond to a marginally stable spacetime closure, hypersensitive to perturbation; a large baseline would prevent long-lived structure from ever forming.

The observed small but nonzero value instead indicates near-critical stability: spacetime is sufficiently robust to support nested closures — energy, particles, galaxies, chemistry, and observers — yet not exactly scale-invariant. Cosmic acceleration is therefore not evidence for an unknown component of the universe, but the observational signature that spacetime itself has settled into a stable, but not trivial, fixed-point regime. In this sense, cosmological data are measuring not what fills spacetime, but how spacetime survives its own coarse-graining at the largest scales.

Cosmic Acceleration Data → IIP–VGF Interpretation

Type Ia Supernovae (SN Ia)

- Standard interpretation: Luminosity distances $d_L(z)$ indicate late-time acceleration.
- VGF interpretation: Direct probes of the global baseline stability of the spacetime closure; deviations from decelerating expansion signal a nonzero stabilisation term.

Baryon Acoustic Oscillations (BAO)

- Standard interpretation: A standard ruler constraining $D_A(z)$ and H(z).
- VGF interpretation: Precision mapping of how the spacetime shell scales under coarse-graining; tests whether the closure is at, or near, a fixed point.

Cosmic Microwave Background (CMB)

- Standard interpretation: Early-universe anchor for cosmological parameters.
- VGF interpretation: Boundary condition fixing the initial stabilisation regime from which the spacetime closure evolves toward its late-time equilibrium.

Equation of state $w(z)$

- Standard interpretation: Parameterisation of dark energy dynamics.
- VGF interpretation: A diagnostic of how tightly the spacetime closure is locked to its fixed point; $w = -1$ corresponds to exact equilibrium, while small deviations indicate slow relaxation.

So in the IIP–VGF framework, acceleration data do not reveal a new substance in spacetime; they reveal how close spacetime itself lies to a stable fixed point of the generative process. That closeness is not accidental.

Cosmic acceleration appears as a natural consequence of the same principle that governs all emergence in the framework: survival under iteration. Only a spacetime closure that is sufficiently stable to persist, yet sufficiently flexible to accommodate nested structures, can support long-lived energy distributions, particles, chemistry, galaxies, and observers. A universe with no baseline stabilisation would be marginal and fragile; one with excessive stabilisation would suppress structure before it could form. The observed acceleration therefore reflects neither fine-tuning nor design, but selection by persistence: spacetime has settled into a regime that survives its own coarse-graining while remaining hospitable to higher-order closures.

Cosmic acceleration is thus not an anomaly to be explained away, but the large-scale signature that spacetime itself satisfies the conditions required for the world to be as we find it — because only such a world can endure.

In other words, cosmic acceleration as we find it is not evidence for something extra in the universe; it is literally the objective evidence that **spacetime itself has survived**.

www.ingramcontent.com/pod-product-compliance
Lightning Source LLC
LaVergne TN
LVHW060044210926
848481LV00013B/3949
9781917403245